the lenin era
1900-1924

by Stuart A. Kallen

Consultant: Margaret Robinson Preska, Ph.D. Russian History
President, Mankato State University [1979-1992]

Published by Abdo & Daughters, 6535 Cecilia Circle, Edina, Minnesota 55439.

Copyright © 1992 by Abdo Consulting Group, Inc., Pentagon Tower, P.O. Box 36036, Minneapolis, Minnesota 55435. International copyrights reserved in all counties. No part of this book may be reproduced in any form without written permission from the publisher. Printed in the United States.

Photo credits: Archive Photos-cover, 9, 31
FPG International-5, 6, 15, 17, 18, 20, 23, 28, 36, 43
Globe Photos-38, 40
UPI/Bettmann-12, 25, 32, 35, 44

Edited by: Rosemary Wallner

Library of Congress Cataloging-in-Publication Data

Kallen, Stuart A., 1955-
    The Lenin era / written by Stuart A. Kallen ; [edited by Rosemary Wallner].
      p. cm. — (The Rise & fall of the Soviet Union)
    Includes index.
    Summary: Examines the events in the Soviet Union during the era of Lenin and the aftermath of his death.
    ISBN 1-56239-101-1 (lib. bdg.)
    1. Soviet Union—History—1917-1991—Juvenile literature. 2. Lenin, Vladimir Il'ich, 1870-1924—Juvenile literature. [1. Soviet Union—History—Revolution, 1917-1921. 2. Lenin, Vladimir Il'ich, 1870-1924.] I. Wallner, Rosemary, 1964-   . II. Title. III. Series: Kallen, Stuart A., 1955- Rise & fall of the Soviet Union.
DK266.K323   1992
947.084'1—dc20
                                                                                             92-13473
                                                                                                CIP
                                                                                                   AC

table of contents

Page

- 4   may 1887
- 8   vladimir becomes lenin
- 11   the revolutionary cause
- 16   war and peace and revolution
- 21   lenin's triumphant return
- 24   riots, rumors, and lies
- 27   the october revolution
- 30   reality sets in
- 39   red versus white
- 42   the bitter end
- 46   glossary
- 48   index

may 1887

Dawn broke over the Russian countryside. Behind the foreboding stone walls of a prison, a convict climbed out of his hard, iron bed. Several soldiers stood outside of his cell, waiting to take him to his doom. The prisoner, a college student from St. Petersburg, had been convicted of plotting to kill Czar Alexander III. He had been followed by secret police for weeks. They finally caught the prisoner near the Czar's carriage with a hollowed out book disguising a bomb.

With his hands tied tightly behind his back, the prisoner was marched to the gallows. In the cold, grey light of that morning, the prisoner, Alexander Ulyanov (You-LYE-ah-nov), mounted the stairs to the scaffold. As the hangman placed the noose over his head, Ulyanov cried out, "The people will prevail, long live the Revolution!" The trap doors opened, and the revolutionary swung in the air, his last breath gone with the wind.

Alexander III (second from left) visits Berlin in 1889.

Karl Marx (1818-1883)

When news of the hanging reached Ulyanov's family, they reacted with tears. But while the rest of the family cried, Ulyanov's younger brother Vladimir made a vow. One day he too would become a revolutionary like his brother. He would overthrow the Czar and bring justice to Mother Russia, the land that he loved. Vladimir had studied the words of Karl Marx, a German philosopher. He believed Marx when he said that the workers of the world would unite and throw off the chains of slavery that were placed on them by the rich and powerful. The hanging of Alexander Ulyanov would change his little brother, Vladimir, forever. And because of the change that came over Vladimir, the course of world history would be altered forever, affecting almost everyone on the earth.

Vladimir had been born in Simbirsk, Russia, in 1870. His father was a hard working, stern, school inspector. His mother was the daughter of a medical doctor. Although neither of Vladimir's parents took an interest in politics, all five of his brothers were revolutionaries. Vladimir enjoyed a rather pleasant childhood — the same could not be said for millions of other Russians.

vladimir becomes lenin

The later years of the nineteenth century were difficult for Russia. Czar Alexander II had freed the peasants from slavery but most of them were still tied to the land in conditions not much better than before. Cruel punishments awaited those who could not earn enough to pay the high rents expected by wealthy landlords. Revolutionaries made headlines with bombs, torture, and assassinations, while the government repressed, censored, and imprisoned thousands. Secret police lurked in every school, town, and village, arresting anyone who complained.

Russia, the largest country in the world, is rich in natural resources like gold, oil, and minerals. Unfortunately, because of the extremely frigid weather and old-fashioned systems of production, Vladimir's Russia at the end of the nineteenth century was a hard place to live where only the strong survived.

Vladimir Lenin (1870-1924)

Shortly after his brother was hanged, Vladimir Ilyich Ulyanov began studying law at Kazan University. Somewhere during the course of the year he joined with college revolutionaries and was expelled from Kazan. By 1893, Vladimir was living in St. Petersburg where he was one of the most active leaders of the Marxist groups there.

After traveling to Switzerland in 1895 to meet and study with other Marxists, Vladimir returned to Russia. After organizing the scattered Marxist movements into a single party, Vladimir was arrested and sent to prison. In February 1897, as part of his punishment, Vladimir was sent to the frozen lands of Siberia for three years. Vladimir used his exile in Siberia to his advantage. He rented a room in a peasant's house and spent his days exchanging letters with other revolutionaries in Russia and abroad. During this time, Vladimir started using the name Lenin. His writings were read in underground circles and soon he became known as Lenin by Russia's revolutionaries and secret police alike.

the revolutionary cause

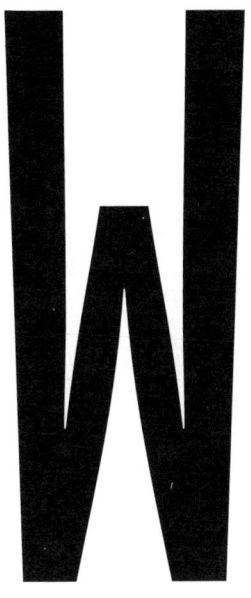

While Lenin was in exile, three years of crop failures caused starvation on a massive scale. In 1897, 1898, and 1901, millions of peasants were forced off the land when their crops failed to grow in the scorching, rainless Russian summer. Thousands of people died of starvation and disease as Russia's Financial Minister traded scarce food overseas for gold. This brought peasants swarming into the cities. People protested the intolerable conditions by striking and demonstrating in the street. The number of people arrested for political crimes reached new peaks as secret police rounded up anyone remotely involved with protest.

After Lenin left Siberia, he spent most of the following years traveling and studying in Europe, living mostly in England and Switzerland until the spring of 1917. Historians say that Lenin was absolutely devoted to the Communist Marxist cause. People who disagreed with him were dealt with harshly.

Hungry children in Russia; where even the smallest bits of food brought smiles to the poor, starving people.

To Lenin, any degree of lying, force, or deceit was allowed, as long as it helped his cause. This ruthlessness in thought and action became the mark of Communist leaders for generations.

While Lenin lived in Switzerland, he worked tirelessly and with great energy to build up the Communist party. He worked with full-time professional revolutionaries who were strictly disciplined and entirely under his orders. Through criminal activities such as robbing banks and stealing from the rich, Lenin financed his party. These activities caused division within the Communist party and many bitter battles were fought by people who thought Lenin a scoundrel and a dictator.

Meanwhile in Russia, most people were interested in another kind of revolution. The Industrial Revolution had finally reached Russia, and factories sprung up in town after town. Russia's vast untapped resources slowly improved the lives of its impoverished people. Russia became the fifth largest industrial power in the world. There was a huge increase in the production in steel and rubber, fueled by the unending demand for galoshes — rubber boots needed to survive the Russian winter.

Then World War I broke out in 1914, causing a huge upsurge of patriotic feeling. Peasants and workers rallied to defend the Russian Motherland from German attack. Very few wished to listen to Lenin and his tiny band of followers in Switzerland.

German troops pass through a Russian town during

war and peace and revolution

By the beginning of 1917, the war with Germany had turned into a bloody disaster. Russian troops deserted by the tens-of-thousands. Even the closest friends of Czar Nicholas II began to think that he was unfit to lead the country.

Anti-German hysteria ran high. Even the name of St. Petersburg was changed because it sounded too German. The city's name was changed to Petrograd. Bread rationing was ordered, causing a panic in the streets.

By the third winter of the war, workers had become tired of the war conditions. Prices had risen sharply while wages had not. There were long, weary lines at every food shop. Hundreds-of-thousands of workers marched in the streets. Factories producing much needed war supplies were closed by strikes. Soldiers refused to take action against the protesters, and many actually joined ranks with the demonstrators.

The first hours of the Russian Revolution.

Czar Nicholas II; even his closest friends thought he was unfit to run the country.

Vast crowds swelled the streets of Petrograd shouting revolutionary slogans. Food stores were looted, a law court was burned and a huge crowd forced its way into the prison, releasing the prisoners and setting the building on fire.

Nobody knew what would happen next because most of the revolutionary leaders, including Lenin, were in prison or out of the country. By March 14, 1917, Czar Nicholas resigned, ending almost four hundred years of Czarist rule. Soon, Nicholas was exiled to Siberia. Within a year, he and his entire family were murdered.

Alexander Kerensky headed Russia for the four
months after the abdication of the Czar

lenin's triumphant return

After the Czar's government was toppled, a young Socialist leader named Alexander Kerensky tried to reform Russia's chaotic government. With his handsome features and fiery manner of speech, Kerensky attracted a large group of followers. Lenin had been living in Berne, Switzerland, and when he heard about the fall of the Czar, he hurried home to Russia. At stopovers on the long train ride home, Lenin was surprised by large groups of workers chanting "Lenin! Lenin!" His train pulled into Petrograd station on April 16, 1917, six days short of his forty-seventh birthday. A reception befitting a conquering hero awaited him. Crowds chanted, bands played, and spotlights swept the sky. Lenin climbed up on the hood of an armored car and whipped the crowd into a frenzy with passionate slogans.

As Lenin approached the newly seized palace, teeming throngs of people awaited him at every intersection.

All through the night, Lenin shouted out speeches to cheering party leaders and workers. When the cheering was over, Lenin set about changing Russia quickly and radically. His Bolshevik party would be called the Communists and government ministers would be called commissars. He set up regional people's councils called soviets, that would draft new laws. Lenin's favorite slogan became, "All power to the soviets." Lenin told peasants to seize the land from the landowners and called for an end to the war with Germany.

Lenin giving his speech in Petrograd, 1917.

riots, rumors, and lies

All of Lenin's speeches and pronouncements could not hide the fact that there were eighteen different revolutionary groups fighting for Russian hearts and minds. Lenin thought there was too much arguing going on and planned for Kerensky's downfall. Bolshevik propaganda spewed forth from Lenin's pen. Inflammatory posters urging people to burn the palaces and kill the rich appeared on city walls. Crowds of workers and soldiers roamed the streets of Petrograd venting their anger in bloody clashes.

Kerensky fought back by imprisoning Leon Trotsky, a close friend of Lenin. Forged documents proving Lenin to be a German spy were circulated. Orders were issued to arrest Lenin. The government blamed him for inciting riots and planning a revolution. Worst of all, they said, he was a foreign agent, an enemy of the state. Lenin became a fugitive, using wigs and disguises to hide from authorities.

Trotsky and Lenin during a parade of Russian troops.

He became a master at hiding secret papers in the soles of his shoes, in false bottomed trunks, and fake books. Rumors swirled through the country, saying Lenin had escaped in a German submarine. In fact, he had shaved off his beard and lived in a small town in a hut near a lake. His desk and chair were two tree stumps and he was forced to live on a diet of tea and potatoes roasted over a campfire. After a time, Lenin disguised himself as a stoker on a railroad train. With the help of sympathetic railroad workers, Lenin slipped past heavily guarded border checkpoints into Finland. There he found work as a farmer.

the october revolution

On October 26, 1917, the revolutionary flag was raised over the palace of the Czars. By that time, Lenin had snuck back into the country and was stirring up revolutionary fervor through secret communications and illegal newspapers. Weeks before, a conservative general had tried to topple Kerensky's government. His soldiers refused to fire on their "brothers," and many of them defected to Trotsky's Bolshevik Red Guard militia. Lenin, again in disguise, took over leadership of the Bolshevik army, taking control of the telephones, newspapers, railway stations, and power plants. The isolated government of Kerensky was now cut off from calling up soldiers in other parts of the country. Fifty thousand Red Guards descended on the palace. Kerensky left the palace in disguise, and escaped in a fast car provided by the United States embassy. He eventually made his way to the United States, where he died in New York City in 1970 at the age of eighty.

Street fighting in St. Petersburg, 1917.

Lenin made himself the head of the Russian government. He made banks the property of the state, called once again for peace with Germany, and abolished private land ownership. Suddenly, Communism became the focus of fear by the capitalist nations of the world. Outside of Russia, Lenin became one of the most feared and hated men in the world.

reality sets in

Soon after seizing power, Lenin realized that actually building a new Russia would be a lot harder than just talking about it. On July 10, 1918, a new Russian constitution appeared. Lenin decreed new and radical changes. With the stroke of his pen, he put every bank, mine, industry, and farm under the ownership and control of the national government. All the property of the Russian Orthodox Church was put under government control. Within ten years, three-quarters of Russia's churches would be boarded up, destroyed, or used for grain storage. Lenin's new constitution also stated that national groups and minorities would be granted new rights, work days would be shortened to eight hours, education would be free, and women would have equal rights with men. Red became the national color and the Red Army was strengthened under the skillful tactics of Trotsky. All men were drafted into the Army at the age of eighteen.

Lenin speaks on the virtues of Communism to crowds in Moscow's Red Square, 1918.

Stalingrad was compared to a hell, with its 200 days of fires and bomb blasts.

Almost as soon as Lenin seized power, troubles began to brew in the countryside. German troops were practically on the outskirts of Petrograd as World War I raged on. The Ukrainians wanted their independence and rebelled. Cossacks and other military men threatened a counterrevolution. Anti-Bolshevik army units began to harass people who were loyal to Lenin. This army, called the White Army, was bitterly opposed to the Red Army. Britain, France, and the United States gave men and supplies to the newly formed "Whites."

Many of the Whites were former government officials, nobles, military men, and clergy. They referred to Lenin and his new system as the "Red Menace." Meanwhile, Lenin made peace with Germany and ended Russia's involvement in World War I. The Germans made Lenin pay dearly, humiliating him by taking one-third of Russia's population and one-fourth of its land, including valuable farming and industrial areas. Lenin could only hope that the revolution would spread to Germany and Russia would regain their lost lands.

Adding to this chaos, Lenin decided to move the Soviet capital from Petrograd to Moscow, in part to escape the entrenched German army. The move was not a popular one, and Lenin and his co-workers had to sneak out of Petrograd under the cover of darkness. The Communist headquarters was moved into the run-down buildings of the Kremlin, a walled in citadel from which Czars had ruled centuries before. Lenin took the role as Supreme Soviet Leader and tried to bring peace to his embattled country.

Just as the war with Germany was ending, the war within Russia was getting a stranglehold on its people. Anarchy, starvation, terrorism, and despair held everyone captive. Between 1918 and 1920, more than seven million Russians died of starvation. After twenty years of wars and revolutions, Russia's production of goods and food had sunk to new lows. Simple people were terrorized in cities and the countryside by bands of robbers, army deserters, and former prisoners of war. Some people were so hungry that they turned into cannibals.

Hungry peasant refugees arrive in Moscow, 1921.

Some of the many German prisoners taken by the Russians in 1918.

Meanwhile, Lenin started a campaign of terror to solidify his power. He ordered large numbers of political enemies imprisoned and started a brutal and ruthless secret police force called the Cheka (Cheka stood for the "Extraordinary Commission to Combat Counter-Revolution and Sabotage"). The Cheka went on a rampage, arresting and murdering thousands of "counterrevolutionaries." Tens of thousands of people were sent to concentration camps, labor camps, and mines.

Lenin addresses the troops before their departure to the front, May 1920.

red versus white

By the summer of 1918, a savage civil war gripped Russia. For three years, bloody battles were fought between the Reds and the Whites. Soldiers from Czechoslovakia and Poland joined in the fray. Crazed soldiers looted and burned hundreds of towns. The lofty ideals of Lenin were put on the back burner as he labored day and night to save Russia from complete collapse.

In August 1918, the head of the Petrograd Cheka was killed, and Lenin himself was seriously injured by two bullets fired from the gun of a young woman named Fanya Kaplan. The young woman, who had a history of mental illness, declared Lenin had betrayed the revolution. Kaplan was arrested and executed four days later. The incident triggered a massive hunt for other counterrevolutionaries and more than one thousand people were executed. Lenin lived with that bullet lodged in his neck until the day he died.

Lenin and his wife with their nephew and his friend, 1922

By 1921, the civil war was over but the country was in ruins. The White Armies were crushed and all other foreign armies left Russian soil. Food, relief aid, and tons of supplies started pouring into Russia from foreign countries, including the United States. Experts from all over the world descended on Russia to help get the country back on its feet. Lenin realized that the words of Karl Marx rang hollow when put to the test of reality. A new economic policy was instituted, combining Communist doctrine with features of capitalism. Private buying and selling of goods were permitted and taxes were collected. Farms were reorganized, electric plants were built and industrial production grew. People's morale improved when food and goods became available.

the bitter end

After Lenin was shot, his health continued to deteriorate. In May 1922, he suffered the first of two strokes that would afflict him that year. In 1923, Lenin suffered another stroke that paralyzed the right side of his body and resulted in his loss of speech. Throughout his time of illness, Lenin fought the inevitable with stubborn strength. With the help of his wife and sister he could walk. He even taught himself to write again with his left hand. Lenin's involvement in running the government was restricted to visits by top Kremlin leaders. These leaders included Trotsky and a new, powerful leader named Joseph Stalin.

As Lenin realized that death was near, he thought about who would take his place. Stalin, during the time of Lenin's sickness had been skillfully concentrating himself in a position of great power. Lenin greatly distrusted Stalin, and Trotsky despised him. Both men thought that Stalin was a simple peasant with an inferior intellect.

Stalin and Lenin in Moscow, 1922

Lenin in his wheelchair, shortly before his death.

Even though Lenin had arrested, tortured, and murdered thousands of people, he thought Stalin was too vicious to lead Russia.

On the evening of January 21, 1924, Lenin died. The time was 6:50 p.m. and the temperature outside was thirty degrees below zero. In keeping with Russian tradition, all the mirrors in the house were covered at once, and every clock in his house was stopped at 6:50. They have remained that way ever since. Lenin was fifty-four years old.

Six days later, in the bitter Russian cold, hundreds of thousands of people lined up to view Lenin's body and to pay their last respects. Moscow was shut down so mourners could attend the revolutionary's funeral. Lenin's embalmed body has been on display in a tomb on Red Square, just outside the Kremlin walls, ever since. Today there is talk of returning Lenin's body to his home town so that he can finally be laid to rest beside the bodies of his family. After nearly seventy years, the revolutionary might go home again.

glossary

cannibal — A person who eats human flesh.

capitalism — An economic system where individuals own goods and where prices, production, and distribution are determined by competition in a free market. The United States is a capitalist country.

Cheka — The secret police force in the Soviet Union from 1917 until 1954. In the Russian language, Cheka stands for the "Extraordinary Commission to Combat Counter-Revolution and Sabotage"

Commissar — The head of a government department in Russia.

Communism — A system of government based on common ownership of property. In Soviet Communism all economic and social activity is controlled by a powerful central government.

Communist — A person who believes in Communism.

cossack — An elite group of soldiers from Southern Russia whose specialty was fighting on horseback. Organized as cavalry in the czarist army.

counterrevolutionary — A person who tries to start a revolution against a government that has just had a revolution.

czar — Title given to the kings of Russia. After the 1917 Revolution, Russia was no longer ruled by czars.

Kremlin — A fortified building in Moscow that contains the government center of the Soviet Union. Also the name used for the entire Soviet government.

militia — A group of civilians enrolled in a country's armed forces. Militias are only used during emergencies.

minority — A smaller part of a population with a different race, religion, or ethnic background. Many times minorities are treated differently, often badly.

propaganda — The repetition of ideas, information, or rumor in books, newspapers, or the media. Propaganda is repeated with the purpose of helping or hurting an institution, a cause, or a person.

Soviet Union — The common name used for the Union of Soviet Socialist Republics or USSR. Russia was renamed the USSR after the 1917 Communist revolution.

index

Alexander II-8
Alexander III-4,5
Bolsheviks-22,24,27
Cheka, the- 34,37,39
Cossacks-33
Industrial Revolution-13
Kaplan, Fanya- 39
Kerensky, Alexander-20,21,24,27
Marx, Karl-6,7,41
Nicholas II-16,18,19
October Revolution, the-27
Petrograd- 16,21,23,33,34,39
Red Guards-27,30,39
"Red Menace," the-33
Russian Revolution-15,16
St. Petersburg-16,28
Stalin, Joseph-42,44,45
Trotsky, Leon-24,25,27,30,42
Ulyanov, Alexander-4,7
Ulyanov, Vladimir (Lenin)-7-9,11,13,14,19, 21-27,29-31,33,34,38-45
Whites, the-33,39,41

MAY 1 5 1995

$13.99

```
J
947.081    Kallen, Stuart A.
K          The Lenin era
```

```
J
947.081    Kallen, Stuart A.
K  2530526 The Lenin era
```

Peninsula Public Library

280 Central Avenue, Lawrence, N.Y.

Phone: CE 9-3262

PN

GAYLORD M